INVERTEBRATES

Animal Group Science Book For Kids

Children's Zoology Books Edition

SPEEDY PUBLISHING

Speedy Publishing LLC
40 E. Main St. #1156
Newark, DE 19711
www.speedypublishing.com

There are likely millions of invertebrates living in your house right now. They are called dust mites

Invertebrates are animals that do not have backbones, also called vertebrae. Invertebrates are cold-blooded.

Invertebrates are the most diverse species on Earth. The majority of animal species are invertebrates; one estimate puts the figure at 97%.

Invertebrates are found just about everywhere in both terrestrial and aquatic habitats.

There are a wide variety of interesting ocean animals that are invertebrates.

Invertebrates occupy more than 30 groups of animals, these include sponges, mollusks, crustaceans, worms, insects, spiders, centipedes and etc.

Mollusks have a soft body that is covered by an outer layer called a mantle. Many mollusks live inside a shell.

Crustaceans form a very large group of arthropods, crustaceans have an exoskeleton, which they moult to grow.

Most crustaceans are free-living aquatic animals.

The body of a crustacean is composed of body segments, which are grouped into three regions: the head, the thorax and the abdomen.

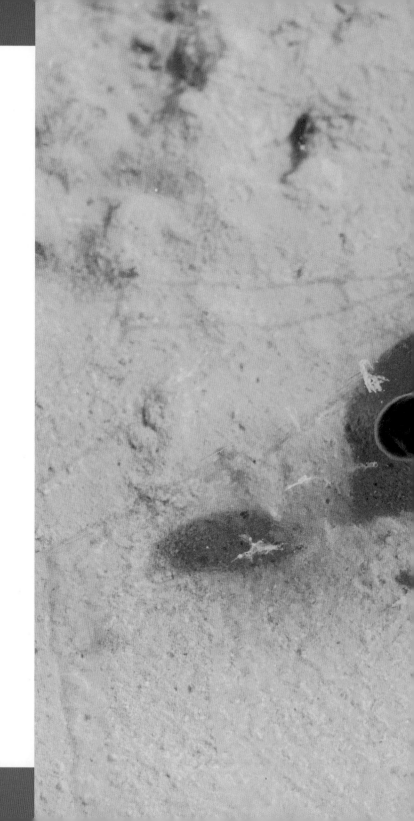

Worms describe many different distantly related animals that typically have a long cylindrical tube-like body and no limbs.

Worms live in almost all parts of the world including marine, freshwater, and terrestrial habitats.

Insects are among the most diverse groups of animals on the planet. There are over 1 million species of insects.

Insects may
be found
in nearly all
environments,
although
only a small
number of
species reside
in the oceans.

Insect growth is constrained by the inelastic exoskeleton and development involves a series of molts.

The insect's head has a pair of antennae, and a pair of compound eyes.

Spiders are
air-breathing
arthropods
that have
eight legs and
fangs that
inject venom.

Centipedes are carnivores which eat insects and worms. They have a poisonous bite to help them kill their prey.

Scorpions have been around for millions of years and they live in some of the harshest places on earth.

Printed in Poland
by Amazon Fulfillment
Poland Sp. z o.o., Wrocław